LIFE SCIENCE IN DEPTH

FEEDING RELATIONSHIPS

Ann Fullick

Heinemann
LIBRARY

www.heinemann.co.uk/library
Visit our website to find out more information about Heinemann Library books.

To order:

 Phone 44 (0) 1865 888066

 Send a fax to 44 (0) 1865 314091

Visit the Heinemann bookshop at www.heinemann.co.uk/library to browse our catalogue and order online.

First published in Great Britain by
Heinemann Library, Halley Court, Jordan Hill,
Oxford OX2 8EJ, part of Harcourt Education.

Heinemann is a registered trademark of
Harcourt Education Ltd.

Editorial: Sarah Shannon and Dave Harris
Design: Richard Parker and Q2A Solutions
Illustrations: Q2A Solutions
Picture Research: Natalie Gray
Production: Chloe Bloom

Originated by Modern Age Repro
Printed and bound in China by South China
Printing Company

10 digit ISBN: 0 431 10898 6 (hardback)
13 digit ISBN: 978 0 431 10898 8

10 digit ISBN: 0 431 10905 2 (paperback)
13 digit ISBN: 978 0 431 10905 3

11 10 09 08 07
10 9 8 7 6 5 4 3 2 1

British Library Cataloguing in Publication Data
Fullick, Ann, 1956-
  Feeding relationships.
  - (Life science in depth)
  591.5'3
A full catalogue record for this book is available
from the British Library.

Acknowledgements
The publishers would like to thank the following
for permission to reproduce photographs:
Alamy pp. **40** (Dennis Kunkel), **21** (J.T. Fowler),
38 (Kelvin Aitken), **12**, **57** (Nigel Cattlin),
22 (Ulrich Doering); Corbis pp. **8** (Bettmann),
6 (BSPI), **35** (David Muench), **25** (Jeffrey
Rotman), **13** (Johnathan Smith/Cordaiy Photo
Library), **51** (Joseph Sohm), **19** (Lester V.
Bergman), **50** (Martin Harvey), **54** (Michael
Yamashita), **31** (Phil Schermeister), **47** (Photo
Library international), **28** (Tom Brakefield),
17 (Wolfgang/Kaehler), **26**; Getty Images p. **7**
(PhotoDisc); Harcourt Education Ltd p. **10**
(Peter Gould); Lonely Planet Images p. **58**
(Sara Jane Cleland); Pacific Ecoinformatics and
Computational Ecology Lab p. **43** (FoodWeb3D,
written by Rich Williams, www.foodwebs.org);
Rex Features pp. **1**, **5** (Richard Austin); Science
Photo Library pp. **29** (Eye of Science), **15**;
Anthony Blake p. **30** (Tim Hill).

Cover photograph of a venus fly trap, reproduced
with permission of Corbis/David Aubrey.

Our thanks to Emma Leatherbarrow for her
assistance in the preparation of this book.

Contents

Words printed in the text in bold, **like this**, are explained in the Glossary.

Everything needs food

Wherever you are, whatever you are doing, you will be surrounded by living things. From the **bacteria** in the air to the plants in the garden, the birds in the sky, and the people you know, life is all around you – and **energy** is the key to life. Living things need energy to move, to grow, to reproduce, to power their sense organs – to live! All this energy must come from somewhere, and it comes from food. Food is broken down to provide energy for **cells**.

WHERE DOES THE ENERGY COME FROM?

All living things need food for energy – but plants do not eat. This is because green plants can make their own food in a special process called **photosynthesis** using:

- carbon dioxide, a gas in the air.
- water, which they get through their roots from the soil.
- light from the Sun.

Plants can then use the energy from the food they have made to keep their cells alive. Animals are not quite so lucky. They cannot make their own food, so to get the energy they need to live and grow, they have to eat either plants or other animals. They have to overcome all the problems of finding enough food and – if they eat other animals – catching the other animal first. Even when an animal has eaten, there are still problems. The food needs to be broken down into small **molecules**, which can be used in the cells to provide energy.

THE LINKS OF LIFE

Imagine this scene. A meadow sparkles in the early morning sunlight. Hidden in the grass, a tiny field mouse sits nibbling some grass seeds when, in the blink of an eye, a hawk swoops down and snatches it up before flying off to feed. These living things – the grass, the mouse, and the hawk – are all linked together in a **food chain** by what they eat. The grass makes food using the sunlight. The mouse eats the plant to get food it cannot make itself, and the hawk eats the mouse for the same reason.

Food chains like this are simple but real life is not. Field mice do not just eat grass seeds. Hawks also eat rabbits and lizards as well as mice. In any one place, different food chains join together to make complex **food webs**. You will find out more about these different feeding relationships later in the book.

Living creatures like this mouse are linked to many other plants and animals through food chains and food webs.

Did you know..?

The waste gas made in the process of photosynthesis is oxygen, which all living things need to use to get the energy from their food. So photosynthesis is the powerhouse of the planet!

The start of the chain

All animals need to eat things, so the world is full of feeding relationships between plants and animals. These are known as food chains and food webs. At the start of most feeding relationships you will find a plant. This is because plants can make their own food by photosynthesis, and this food is then used by almost every other living **organism** on the Earth in one way or another. In food chains and webs, plants are known as the **producers** because they use light, carbon dioxide, and water to produce the food that everything else relies on.

LEAVES – THE FOOD FACTORY

For photosynthesis to work, a plant needs carbon dioxide, water, and sunlight. It also needs a way of carrying the sugar molecules that are made all around the plant, and a way of getting rid of the waste oxygen gas that is produced.

The leaves on a tree, like many other plants, are arranged to make sure they get as much light as possible. They act as factories making food for the plant – and feeding the rest of the living world as well.

Photosynthesis takes place in the green parts of a plant, especially the leaves. This is because the leaves are full of a special green chemical called **chlorophyll**. Chlorophyll traps the energy from sunlight falling on the leaves, ready to be used to make food.

If you lie under a tree and look up at the sky you will see just how well organized plants are. The leaves of a plant are often flat, to catch as much sunlight as possible. They are usually arranged so that they all get some light. What is more, the leaves are thin, so gases can pass through them easily, and they have a good network of veins for carrying food and water within the plant. The leaves of a plant are perfect production lines for making food on a big scale!

Leaves come in a range of different shapes and sizes. The leaves of ferns are known as fronds. They form an interesting pattern that you can see in this picture. The fronds unfurl into this pattern as they grow.

THE REST OF THE INGREDIENTS

Leaves are very important for plants because without them a plant cannot photosynthesize properly. But it is no good just having leaves and a supply of light. Plants need carbon dioxide and water as well if they are going to make food.

Carbon dioxide is found in the air around us. It is poisonous to animals and people in large quantities, but it only makes up a small part of the air (about 0.04 percent). Plants need carbon dioxide. It gets into the leaves of a plant (along with the rest of the air) through tiny holes called stomata.

Water is the other vital ingredient for photosynthesis. Plants get water from the soil, taking it up through their roots. Plants have a big system of veins running all through their stems, roots, and leaves. These veins carry the water from the roots and up to the leaves, ready for use in photosynthesis.

SCIENCE PIONEERS Melvin Calvin

Melvin Calvin was born in St. Paul, Minnesota, USA in April 1911. He began his famous work on photosynthesis in 1945. Calvin set up a team of young scientists and had the ground-breaking idea of using people from lots of areas of science – biologists, chemists, and physicists – to work together, bringing different skills to the research.

Melvin Calvin and his team of scientists unravelled the secrets of one of the most important reactions on Earth. In 1961 Calvin won the Nobel Prize for Chemistry for his discovery of the chemical pathways of photosynthesis.

They used a **radioactive** molecule called carbon-14. This showed up in experiments and allows the scientists to follow the reactions of photosynthesis. It was very difficult, because the reactions had to be carried out in living cells. After years of work, Calvin and his team finally mapped the complete path that carbon travels through a plant during photosynthesis. The reactions are called the Calvin Cycle in his honour.

HOW DOES PHOTOSYNTHESIS WORK?

In photosynthesis, sunlight provides the energy to turn carbon dioxide and water into food, similar to the way heat provides the energy in an oven to turn flour and water into bread.

Photosynthesis takes place in the **chloroplasts** in the leaves. In a series of chemical reactions, carbon dioxide and water are joined together. This produces a sugar called **glucose** and oxygen gas. The glucose is carried all around the plant to where it is needed. It is usually turned into starch, a **carbohydrate** that can be easily stored for when the plant needs energy.

Photosynthesis can be summed up in this word equation:

$$\text{CARBON DIOXIDE + WATER} \xrightarrow{\text{LIGHT}} \text{SUGAR + OXYGEN}$$

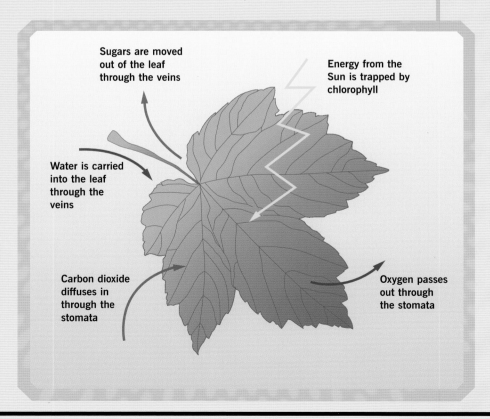

Sugars are moved out of the leaf through the veins

Energy from the Sun is trapped by chlorophyll

Water is carried into the leaf through the veins

Carbon dioxide diffuses in through the stomata

Oxygen passes out through the stomata

WHAT AFFECTS PHOTOSYNTHESIS?

Plants are found all over the world. They even live in the Arctic circle, and survive in deserts. They form the great, lush rainforests of the tropics and the stunted landscape of the **tundra**. Plants play a vital role in feeding the animal population of the world, including all the people, so it really matters how quickly they photosynthesize and grow. Light, carbon dioxide levels, and temperature are known as limiting factors because they all affect the amount of photosynthesis a plant can carry out.

CLASSIC EXPERIMENT Measuring the effect of light on the rate of photosynthesis

In 1754 Charles Bonnet observed that gas bubbles were given off by a leaf underwater that was brightly lit.

In an experiment that is carried out in school laboratories around the world, a light shines on Canadian pondweed or Elodea under water. When photosynthesis takes place, a stream of bubbles of oxygen-rich gas is given off. The rate of bubbling can be counted, or the volume of gas given off can be collected and measured.

When the light is moved away from the plant, the rate of photosynthesis falls and the stream of oxygen bubbles slows down. The low light levels are limiting the rate of photosynthesis. If the light is moved closer (while the water temperature is kept constant) the bubbles appear faster and faster, showing that the rate of photosynthesis has gone up.

Bonnet's observation is the basis of a classic experiment that looks at the effect of light on the rate of photosynthesis.

LIGHT

The most obvious thing that affects photosynthesis is the level of light. If there is plenty of light, lots of photosynthesis can take place, but in the dark or very low light, photosynthesis stops, regardless of the other conditions around the plant. For most plants, the brighter the light, the greater the rate of photosynthesis.

TEMPERATURE

Temperature affects all chemical reactions, including photosynthesis. As the temperature rises the reaction speeds up and the rate of photosynthesis will increase. However, photosynthesis is controlled by **enzymes**. Enzymes are **proteins** which means that they can be badly affected by temperatures above 40°C. If the temperature goes too high, the rate of photosynthesis will fall, as the enzymes controlling it are destroyed.

The rate of photosynthesis increases steadily with a rise in temperature up to a certain point, after which the enzymes are destroyed and the reaction stops completely.

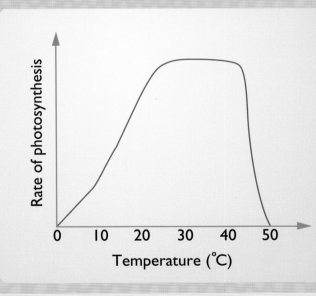

CARBON DIOXIDE LEVELS

The amount of carbon dioxide available also limits how much photosynthesis can take place because without carbon dioxide plants cannot make glucose. In the natural environment of most plants, the carbon dioxide levels in the air are the most common limiting factor when there is plenty of light available.

TAKING CONTROL

In the laboratory, we can look at each limiting factor in turn and see exactly how it limits the rate of photosynthesis. However, for most plants, the balance between the different limiting factors is constantly changing. Early in the morning, low light levels and cold temperature limit the rate of photosynthesis. As the levels of light and the temperature rise, carbon dioxide becomes limiting. On a bright day in winter, the cold temperature probably limits photosynthesis more than a lack of carbon dioxide.

However much we might like to, we cannot affect the rate of photosynthesis of plants growing in the wild, in agricultural fields, or even in our gardens. But if the rate of photosynthesis goes up, crop yields go up as well, so farmers have started to try and take control. They are growing more and more plants in polythene tunnels to increase the temperature. Farmers are also using completely sealed, massive greenhouses, where all of the conditions can be carefully monitored and controlled. The greenhouses contain thousands of plants, along with lots of different sensors. These sensors are linked to computers that continually adjust levels of light, temperature, and carbon dioxide. This makes sure that no factors are limiting, so that the maximum rate of photosynthesis can take place day and night.

Could this be described as force-feeding plants? In greenhouses like this, nothing is left to chance. The plants are even grown in water full of nutrients rather than soil to make sure nothing limits their rate of photosynthesis and growth.

MAKING THE MOST OF THINGS

Plants cannot control their own environments and make sure that they have the right temperature or the ideal levels of light or carbon dioxide. However, plants around the world have developed some amazing **adaptations** that help them to make the most of the conditions they live in.
The leaves on cacti are nothing more than thin spines. This means the cactus doesn't lose much water in the heat of the desert. The spines also stop animals which might try to eat the cactus. What's more, the cacti can photosynthesize very well in their fleshy stems.

Plants like these giant water lilies have leaves that float on top of the water. They can grow to an enormous size, but they need special air sacs inside them to keep them afloat.

Plants that grow in very cold conditions have other problems. These plants need to have an extremely thick waxy layer to act as **insulation**. They also have a special chemical in their cells to protect the cells from freezing – which would destroy them – and let them carry on photosynthesizing in the coldest conditions.

Hairy, rolled, small, or giant leaves, special variations on the chemistry of photosynthesis – plants have tried all sorts of things in the quest to be able to survive in some of the toughest places in the world.

THE ALTERNATIVE PRODUCERS

Most food chains and webs start with a plant producing food from carbon dioxide and water, using energy from the Sun. But in the last thirty years, scientists have discovered some living things that make their own food without relying on light at all.

VENTS IN THE ABYSS

Deep in the oceans, about 2,000–6,000 metres (6,500–20,000 feet) below the surface, is a region known as the abyss. The water there is incredibly cold at about 2–3°C (35–37°F) and it is utterly dark, because no light gets down there at all. The pressure is several hundred times the **atmospheric pressure** on the surface. Down in the abyss, there are vents and chimneys on the sea bed that throw out mineral-rich water full of poisons and gas, all at temperatures of over 350°C (662°F)! The water is so hot because it has been heated by the molten rock within the Earth itself. It is in these extreme conditions that scientists have discovered unique colonies of creatures that rely on each other for food.

EXTREME ORGANISMS

There are no plants at the beginning of these deep-sea food chains because no plants could survive. Instead, scientists have found **chemosynthetic bacteria**. These are bacteria that use energy not from the Sun, but stored in other chemical compounds (particularly hydrogen sulphide), to build new food molecules.

RECENT DEVELOPMENTS Sampling the abyss

In 1998, American professors John Delaney, Deborah Kelly, and John Baross were part of an expedition that, for the very first time, brought parts of some of the chimneys up from the abyss. They sampled the **micro-organisms** and began to **culture** them in the laboratory. Work still continues, and every expedition reveals more **species** that have never been seen before.

These bacteria grow round the vents and chimneys in such large numbers that they form pale mats on the sea bed. Snails graze on the mats of bacteria, and several other organisms can feed off the bacteria too. Some bacteria live in the shells of specialized tube worms, mussels, or clams, where they get protection and minerals in return for providing food. Animals such as strange white crabs then scavenge among the bodies.

These fascinating organisms live at temperatures that would normally destroy body proteins. For many years scientists thought that the deep vent organisms didn't need plants at all. In fact, indirectly, they need plants just as much as any other living things. To get the energy from hydrogen sulphide to make food, the vent bacteria need oxygen. There is plenty of oxygen in the sea, but it all comes originally from the oxygen in the air that was produced by plants during photosynthesis.

Conditions in the abyss are unbelievably hard. This is a "black smoker" on the sea bed which pumps out nutrients. These feed the chemosynthetic bacteria.

Building up the chain

Think about the food you've eaten today. Did you have a sandwich, pizza, dahl, a piece of fruit? Do you know where it came from? Much of the food we eat comes directly from plants. The bread in a sandwich comes from wheat, and salad is obviously plants. The tomatoes, onions and peppers on the pizza are vegetables, while the lentils in dahl are the seeds of a plant, and a piece of fruit is the fruit of a plant!

Even food that comes from animals, such as cheese, meat, butter or milk, also originally comes from plants. The animals that produce our food eat grass, corn, and other plants, and turn that plant material into animal tissues.

In the end, all our food starts off with plants and the process of photosynthesis. All living things are linked by the food that they eat. The simplest links are food chains. Here is a general food chain and some examples:

Plant → animal → animal → other animal

Rice → human

Grass → cow → human

Grass → rabbit → fox

Blackberry → field mouse → hawk

Microscopic water plants → microscopic water animals → small fish → bigger fish → seal → killer whale

SCIENCE PIONEERS Charles Elton

In the 1920s, Charles Elton, a young biologist at Oxford University, travelled to Bear Island off the northern coast of Norway. The island is part of the Arctic tundra, so he made his trips in the summer when there were some plants to observe and the weather was not too cold. Not very many plants could survive in this tough environment, and he wanted to see how the animals shared the small number of plants.

Bear Island only supports a few hardy grasses and small scrubby plants, so it was quite easy for Elton to see which animals fed on the plants – and each other. Arctic foxes were the main large **carnivores**.

Bear Island is a very bleak environment. This is where Charles Elton first watched food chains in action.

During the summer, they fed mainly on birds like **ptarmigan** and sandpipers, that only came for the summer. The birds in turn fed on the leaves and berries of the tundra plants, or on plant-eating insects. Elton described the links between plants, insects, birds, and foxes as a food chain. He also said the first link of a food chain is a plant trapping energy from the Sun by photosynthesis.

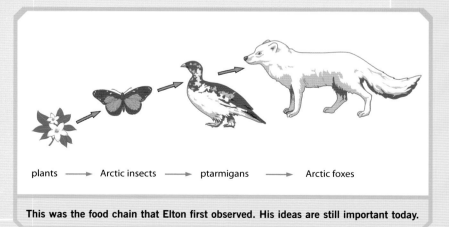

plants ⟶ Arctic insects ⟶ ptarmigans ⟶ Arctic foxes

This was the food chain that Elton first observed. His ideas are still important today.

MORE ABOUT CHAINS

When Charles Elton worked out how food chains worked, he also thought of some names to describe the different links in the chain. These names are still in use today. As we saw earlier, plants are known as producers, because they produce all the new food. Animals that eat plants are known as **primary consumers**, while animals which eat animals that eat plants are called **secondary consumers**, and so on. Just as there are plants at the beginning of every food chain, the end of every food chain has organisms that break down the remains of animals and plants after they have died. These organisms, mainly bacteria and fungi, are known as the **decomposers**. They are not always shown on food chains.

HOW LONG IS A FOOD CHAIN?

Food chains can involve only two types of organism. They rarely involve more than 6 or 7 different types of organism. When humans are involved, the food chains are usually quite short because most human food worldwide involves plants or animals that eat plants, and very few organisms eat humans!

Some food chains are very specialized. The fig wasps that live and die within the flowers and fruits of one particular type of fig tree are part of a very short and specialized chain. On the other hand, you can find examples of the food chain below all over the world:

Plant → insect → small bird → bird of prey

CHAINS IN THE SEA

Some of the most common food chains take place in the two thirds of the surface of the Earth that is covered by seas and oceans. Plants can only live in the top layers of the oceans because that is where light reaches. The plants found there are microscopic – they form a group of organisms called the **phytoplankton**. Although they are small, they can photosynthesize just as well as any land plant. They use up a massive amount of carbon dioxide from the air, and produce a huge amount of plant material.

Fortunately, microscopic organisms known as **zooplankton** eat the phytoplankton around them, so the seas do not get clogged up with tiny plants. In turn, the zooplankton are food for many different species of fish. The smaller fish are then eaten by bigger fish. Sea birds, seals, dolphins, sharks, and whales are just some of the creatures that feed on the bigger fish.

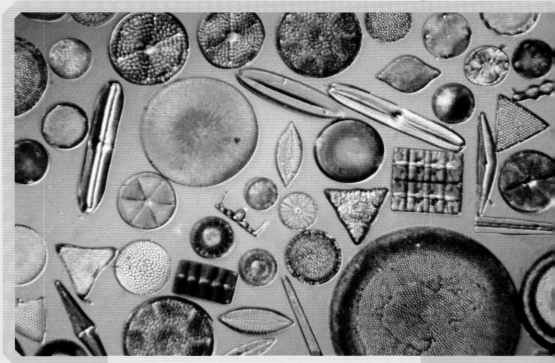

Phytoplankton – small in size but mighty in numbers – produce most of the food needed to support the food chains of the sea.

Did you know..?

The food chains in the sea can be long both in the number of organisms involved and the distances they travel. An ocean food chain can stretch over thousands of miles, as large animals such as fish and whales travel in their search for food.

PRIMARY CONSUMERS

Before the food made by plants can move into a food chain to be used by animals, the animals must be able to eat and **digest** the plants. This isn't as easy as it sounds. Plant cells are surrounded by a tough cell wall made of a chemical called cellulose. This gives the plant strength, but it causes all sorts of digestive problems for many animals. A special digestive enzyme called cellulase is needed to digest cellulose. As most animals do not produce this enzyme, all the useful food material of the plant is locked away inside the plant wall, so the animals have to find other ways to break it down.

Primary consumers that only eat plants are called **herbivores**. A much smaller number of primary consumers can eat both plants and animals – these are called **omnivores**. It is vital to food chains that all primary consumers can actually digest plants. Different organisms have developed different ways of tackling the plant problem. Here are just a few of them.

HERBIVORES IN THE INSECT WORLD

Many insects rely on plants for their food. Some of them overcome the problem of digesting cellulose by avoiding it completely. Butterflies and moths feed mainly on nectar, the sugary liquid made by flowers to lure insects in to **pollinate** them. Nectar contains no cellulose. The butterflies and moths have very special mouthparts with a long, hollow tube called a proboscis which reaches right down into the nectar store and sucks it up. Honeybees have a tubular tongue which they use in the same way.

Aphids are a group of insects that includes greenflies. Aphids also feed on plants, but they have a unique way of feeding. They have very sharp, pointed mouthparts which they stick right into the plant so they can suck up the sugary, food-rich fluid from the plant's veins in the leaves and stems. This means they don't need to digest cellulose.

Termites, on the other hand, like to eat wood, which is mainly made up of cellulose. So they need to have cellulase.

Although termites do not produce the enzyme cellulase themselves, all termites contain **protists** (**protozoa** or microscopic organisms) in their guts which do produce cellulase. The termites bite off small pieces of wood and chew them up; then the protists in their guts break down the chewed mass into sugars, which can easily be used by both the termites and protozoa alike. The protozoa get a protected place to live and food delivered regularly, while the termites get their food broken down for them.

The mouthparts of this Canadian tiger swallowtail butterfly are beautifully adapted for feeding on nectar. When the insect isn't feeding, the delicate proboscis is neatly coiled up to keep it safely out of the way.

Did you know..?

When two different types of organisms live together and both benefit from this relationship, it is called mutualism. Some examples are the protists in termite guts, the bacteria in the rumen of sheep (see page 22), or the fungus and the **algae** in a **lichen**.

THE MAMMALIAN SOLUTION

Lots of mammals are herbivores, yet mammals do not make cellulase at all. There are a number of different ways in which mammals have overcome the difficulties of a plant-only diet.

FIRST THE MOUTH...

Many herbivores have specially adapted mouths which help them to crush and break open tough plant cell walls. Their back teeth, the molars and premolars, are often large, flat, and ridged. Herbivore jaws also usually move from side to side as well as up and down. This makes sure the plant material is thoroughly crushed and ground up by the teeth. Herbivores often spend a long time chewing their food, which also helps to break open the cells.

...THEN THE GUT

Herbivores also need very special guts to help them get enough energy from their food. They often rely on bacteria in their **digestive systems** to make the cellulase they need to break down cellulose. The bacteria get a constant supply of food and a warm, protected environment, while the mammal gets sugar from the cellulose and access to the rest of the plant cell contents.

Animals such as sheep and cows have a special area in their guts where the cellulose-digesting bacteria live, called the rumen. Animals that have this area are called ruminants.

Elephants, the largest living land mammals, feed their enormous bodies on plant material.

The plants that ruminants eat go straight to the rumen where the cellulose is broken down. The partly digested food then goes back to the mouth to be chewed again so that the cellulose walls are completely broken before the food is swallowed again, this time into the abomasum (true stomach). This is known as "chewing the cud".

Other herbivores have different ways of digesting cellulose. In rabbits, the cellulose-digesting bacteria are further along in the **large intestine**. The food goes through the gut once, but it is only partly digested. The rabbit produces moist, bulky **faeces**, which it then eats and redigests. This time most of the nutrients are taken from the food and the rabbit then produces the small dry pellets we recognize as rabbit droppings.

Because digesting cellulose is so difficult, most herbivores have very long guts to give themselves the best chance of getting as many nutrients as possible from their food. And in spite of all these adaptations, most herbivores still struggle to digest a lot of what they eat, so they produce large amounts of very bulky, cellulose-rich faeces.

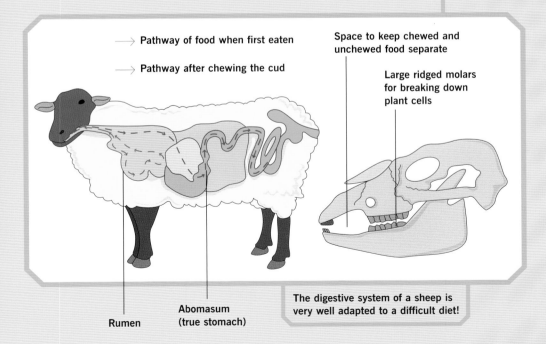

→ Pathway of food when first eaten

→ Pathway after chewing the cud

Space to keep chewed and unchewed food separate

Large ridged molars for breaking down plant cells

Rumen

Abomasum (true stomach)

The digestive system of a sheep is very well adapted to a difficult diet!

The links in the chain

Herbivores eat plants, and they use the energy they gain from digested plant material to build up new animal tissues. However, not all animals in a food chain eat plants – many of them feed on other animals. They are known as carnivores. Carnivores that eat herbivores are known as secondary consumers. However, carnivores that eat other carnivores are known as **tertiary consumers**.

Because carnivores eat other animals, they have a diet that is much easier to digest than that of herbivores. Animals are made up of lots of protein (particularly their muscles and skin), with some fat and carbohydrate, all of which is quite easy to break down in the gut. So carnivores do not have the same difficulty digesting their food that herbivores do. Also, the food they eat is high in energy, so carnivores need to spend a lot less time feeding and digesting than herbivores do. Carnivores spend a lot of their time resting. As a result of their easy-to-digest diet, a carnivore's guts are usually shorter, and they produce relatively little faeces. However, while carnivores do have an easier time digesting their meals, they have to work harder finding or hunting them.

FIRST CATCH YOUR MEAL!

Many carnivores are **predators**, which means they catch and kill other animals. The animals they kill are called **prey** animals. Most animals do not want to be caught and eaten and do everything they can to try and avoid it. So predators have adaptations to help them catch their prey, hold on to it, and kill it. Animals at all levels of the food chain hunt to survive. At the top of every food chain is a predator – and often a fairly large one!

There are many ways in which predators catch their prey. One example is the sea anemone, which look harmless and are often very beautiful, but they are very successful predators. They sit on rocks and wave their coloured tentacles, looking more like exotic plants than animals. When a shrimp brushes against the innocent looking tentacles, tiny cells called **cnidocytes**, also known as nematoblasts, are fired off. Some of these contain poisoned darts that paralyse the shrimp, while others fire off sticky threads that hold the prey as it struggles until the poison takes effect. Then the shrimp is pushed through the mouth of the anemone by the tentacles, to be digested inside the body.

Beautiful but deadly. Sea anemones look like unlikely predators, yet they are very effective killers in their own, specialized way.

Did you know..?

The most ferocious freshwater fish are the piranhas *Serrasalmus* and *Pygocentrus*. These small predators are found in the rivers of South America. A large shoal of piranhas can strip all the flesh off a large animal in minutes, leaving just the bones!

MORE OCEAN PREDATORS

Another predator of the seas is the killer whale, or orca. These black and white members of the dolphin family are about 10 metres (33 feet) long and swim in **pods** containing anywhere between about 5 and 50 whales. They use **echolocation** to judge distances under water with great accuracy. They are also fast and powerful swimmers, and have conical teeth that interlock so they can grip slippery prey such as the squid they love to eat. All of these adaptations make them successful **marine** carnivores.

Killer whales are very good predators, but their numbers are dropping due to a number of problems, including a fall in the populations of the animals they prey on.

WINGED PREDATORS

Many birds hunt and eat other animals. Most of these predators are small birds that eat even smaller insects. They feed on grubs, beetles, worms, or caterpillars in grassland, hedgerows, and forests around the world. Good hearing and the right-shaped beak are the main adaptations needed for hunting these prey. Other birds catch insects in the air. Swifts and swallows are well known for flying fast and low with widely gaping beaks to catch the tiny flying insects that they eat.

KILLER BIRDS

The predatory birds which everyone knows are the great birds of prey. Eagles, buzzards, hawks, kites, owls, and kestrels are just some of the species that are the top carnivores in many food chains. These birds are all adapted to fly in a special way so they can catch their prey. Some have wings and tails that are shaped to allow them to hover in the air while looking for prey. All of them can swoop down fast when they attack their prey. They also have sharp talons on their feet that are adapted for striking and carrying animals. In addition, their curved beaks help them eat their prey.

OTHER FLYING HUNTERS

Not all winged predators are birds. During the night, bats also hunt for food. Bats have a very sophisticated echo-location system (or sonar) which lets them fly safely in the dark without crashing into things. It also makes it possible for carnivorous bats to find the moths they eat. Not all bats eat insects – vampire bats feed on blood from cattle and other large animals.

Did you know..?

The smallest birds of prey are the black-legged falconet from south-east Asia, and the white-fronted falconet from Borneo. They are only about 14–15 centimetres (6 inches) long and weigh about 35 grams (1.2 ounces).

MAMMALS ON THE HUNT

Mammals that are predators range from tiny but ferocious shrews, through to the big cats of Africa and Asia. These include lions, tigers, and cheetahs, and each has their own adaptations. They all have big, strong, sharp teeth to bite into their prey and strong jaws that allow them to hang on. They also have sharp claws and can run very fast in short bursts.

But the tiger hunts alone, while cheetahs sometimes chase their prey in relays, with first one animal and then another chasing the prey at high speed. Lions often hunt as a pride with the whole group of animals surrounding the prey. Success as a predator comes.in many different ways, but the end result is always food to fill a hungry belly.

These lions are typical predators. They have all the adaptations of a top carnivore: big sharp teeth, quick speed, strong claws, and a good hunting strategy.

CARNIVORES BUT NOT PREDATORS

Food chains and food webs contain many animals that are carnivores, but not predators. They do not hunt and kill other animals, they simply feed on them. The scavengers are one of these groups. Scavengers feed on animals that have died naturally, or finish off the remains of an animal killed by a predator. Scavengers such as jackals, hyenas, and vultures pick the bones of a dead animal clean. Sometimes they combine scavenging with killing.

Other animals, such as the bluebottle fly, are also scavengers. The female fly is incredibly sensitive to the smell of death. She often lands on an animal and lays her eggs minutes after the animal has died. The eggs hatch out into **maggots**, which then feed rapidly on the dead animal and grow until they form a **pupa** and emerge as adult flies themselves. In Zimbabwe, maggots and other scavengers can reduce an entire dead elephant to a skeleton in just seven days!

PARASITES – INSIDE AND OUT

There are many animals around the world that live on or in other living animals and feed off them. They are known as parasites. Fleas, ticks, and lice are parasites that live on the outside of an animal and feed on its blood. They have special adaptations so they can survive attempts to kill or get rid of them. For example, fleas can hop large distances, while ticks bury their jaws into the skin so they cannot easily be dislodged.

Other parasites are even more clever. Many of them such as worms or protists live inside the body of their animal hosts. Tapeworms, threadworms, amoeba, and liver flukes all invade a wide range of animals, including humans. Many of them cause diseases such as **malaria** and **sleeping sickness**, that kill and disable millions of people every year.

Tapeworms have hooks and suckers to keep them attached to a person's gut. Their outer coat protects them against the acids in their host's digestive system – and at the same time lets them absorb lots of the digested food!

HAVING IT ALL

Herbivores eat plants and carnivores eat other animals. Their teeth and guts are adapted to the food that they eat. There are also many animals that eat both animals and plants. Often they do so accidentally. When a cow bites off a mouthful of grass, there will inevitably be small insects or other animals caught up on the leaves, and they get digested along the way. In the same way, if a fox catches and eats a rabbit, the rabbit gut will be full of partly digested plants and the plant material will go into the fox. But these animals are still called herbivores and carnivores. It is the animals that deliberately seek out both plant food and that also eat other animals who are the true omnivores – they eat everything.

Human beings and many apes are naturally omnivores, and so are pigs. Being an omnivore creates real problems. The animal food can be digested quite easily, but the plant material cannot. The sort of teeth that are really good for a plant-based diet are not much help in eating meat. So omnivores compromise. They have guts which are a lot longer than a normal carnivore gut, but far shorter than a true herbivore. They have teeth which are sharper and fiercer than herbivore teeth, but with far more scope for crushing and chewing than a true carnivore.

Human beings are omnivores, which means we can eat a very wide variety of foods.

THE END OF THE CHAIN

Living organisms are taking things from the environment all of the time. Plants take minerals from the soil, and these are then passed along food chains into animals. If this was a one-way system, the resources of the Earth would soon be completely used up. Fortunately, the nutrients held in the bodies of plants and animals are all returned to the soil when they die by a group of organisms known as the decomposers.

The decomposers are found at the end of every food chain. They are micro-organisms, such as bacteria and fungi, that feed on droppings and dead organisms. Their waste products return nutrients to the soil. When we say that something has "decayed", it has actually been broken down and digested by the decomposers.

This dead tree is slowly being broken down by the decomposers feeding on it. Without these bacteria and fungi, the world would be covered in the remains of everything that has ever died.

Did you know..?

Decomposers are used in sewage farms to break down the vast amounts of human waste produced every day. With their help, a large sewage farm can treat over 800 million litres (176 million gallons) of waste every day!

The numbers game

The plants and animals living in a similar **habitat** are linked together by food chains. A simple food chain might tell us that rabbits eat grass and foxes eat rabbits. It doesn't tell us how many rabbits there are, or how many foxes. And it certainly doesn't tell us how many blades of grass are needed to feed the rabbits.

When we look a little closer, we can see that it takes thousands of grass plants to feed a small number of rabbits, and these rabbits will only feed one fox. The same is true for lots of food chains. There are lots of producers, providing food for fewer primary consumers who in turn provide food for even fewer secondary consumers.

Using this idea we can build up a simple diagram called a "pyramid of numbers". This shows us the numbers of organisms at each level of the food chain with the producers at the bottom, and the highest level consumer at the top.

grass ⟶ rabbits ⟶ fox

| fox |
| rabbits |
| grass |

plant plankton ⟶ krill ⟶ blue whale

| blue whale |
| krill |
| plant plankton |

Lots of food chains can give us a clear pyramid of numbers.

WHEN NUMBERS DO NOT WORK

The big problem with pyramids of numbers is that they do not work for all the different food chains. Think about an oak tree. One single plant feeds huge numbers of caterpillars and other insects, which in turn feed a smaller number of birds. This makes a very peculiar shaped pyramid.

To overcome this problem, scientists have developed a more reliable way of looking at food chains. Instead of looking at the numbers of organisms involved, they concentrate on the amount of living material, or **biomass**, at each stage of the food chain. It is interesting to see how much of the biomass made by plants actually gets turned into herbivores, and how much of that biomass in turn gets turned into carnivores. The total amount of biomass at each stage in a food chain can be drawn to scale to give a "pyramid of biomass".

Pyramids of biomass give us a much more realistic picture of what is happening in a food chain.

The amount of living material at each stage of a food chain is always less than it was at the previous stage. Not all of the organisms at one stage are eaten by the stage above. What's more, when a herbivore eats a plant, a lot of the biomass from the plant is used to provide energy just for living (moving, growing, reproducing, keeping warm) or is passed out and wasted as droppings. Only the biomass which actually gets turned into herbivore can possibly be passed on to the carnivore at the next level. The same is true at every stage. So a large amount of plant biomass supports a smaller amount of herbivore biomass, which in turn supports an even smaller amount of carnivore biomass.

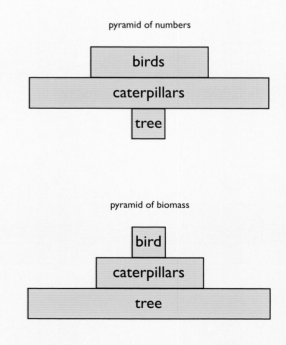

pyramid of numbers

birds

caterpillars

tree

pyramid of biomass

bird

caterpillars

tree

THE FLOW OF ENERGY

Pyramids of biomass give us a much better idea of what is happening in the natural world than pyramids of numbers do, but even they have their limits. For example, if a sample of water from the English Channel is collected and analysed, the biomass of the microscopic animals – the zooplankton – seems to be bigger than the biomass of the tiny plants – the phytoplankton. This shouldn't be possible – there cannot be more animal material than the plants that they feed on!

THE EFFECT OF TIME

The problem is that the sample is taken at a single instant in time. Imagine looking at a garden lawn. The grass looks short and neat as if it produces very little new biomass. But if you go and look at the compost heap where the grass clippings are thrown, you can see that huge amounts of grass have grown. It has just been cut off again regularly. The same is true with the phytoplankton in the English Channel. They reproduce far more rapidly than the zooplankton – but they get eaten quickly too.

Even pyramids of biomass cannot give us all the information we need to understand a complicated system like this. In 1942 an American scientist called Raymond Lindeman decided to look at the total energy flow through an **ecosystem**. He had the idea of looking at ecosystems over time to get a realistic picture of the flow of energy through all the living organisms.

Did you know..?

The fewer stages there are in a food chain, the less energy is lost. Every stage we introduce, such as feeding plants to animals, then eating the animals, lowers the amount of biomass available. This means there is less food and energy to share. In theory at least, if everyone on the Earth ate only plants there would be plenty of food to go around.

SCIENCE PIONEERS Raymond Lindeman

Raymond Lindeman worked at the University of Minnesota in the US. He developed a model to show the way energy flows through the living organisms in an ecosystem, based on the feeding relationships. His ideas are known as the **trophic-dynamic model**. The first **trophic** level contains the plants, the second trophic level is the herbivores, the third trophic level is the carnivores, and so on. Lindeman took into account the way energy is lost through **respiration**, living, and **excretion**. The great advantage of his model is that it looks at a whole ecosystem.

He showed that there is a limit to the number of trophic levels in an ecosystem. There are often three, but rarely more than four or five, because there is a limit to the amount of energy that is passed on from one level to the next. The energy transfer is not very efficient. Usually only about 10 percent of the energy in plants becomes new animal material in herbivores, and then only about 10 percent of the herbivore becomes new carnivore material, and so on. This is because about 90 percent of the energy taken in at each stage is lost through respiration, keeping warm, producing waste, and so on. This explains

Lindeman produced a useful picture of how living organisms interacted over time, which has never failed so far.

why, for example, the killer whale is at the top of the food chain. It has no predators because an animal which was large enough to capture and kill a killer whale would need to spend so much energy hunting that it couldn't realistically eat enough to stay alive! Pyramids of energy are the most complicated but most accurate picture we have of energy flows within an ecosystem.

Food webs

Food chains make sense. The model is easy to understand. You can think of all of the organisms in a habitat taking part in a whole series of different food chains. However, for most situations, this picture is far too simple. Very few animals eat only one type of food – pandas living on bamboo and koala bears which eat only eucalyptus leaves are two rare examples. But most animals eat a variety of different foods. Mice eat grasses, seeds, berries, and roots. Foxes eat mice, rabbits, dead lambs, voles, and beetles. Polar bears eat seals and fish. Animals and plants are linked by complicated interconnecting food webs, which are made up of several different food chains, with some organisms in common.

THE BALANCE OF THE FOOD SYSTEM

In situations where there is a simple food chain, for example the giant panda feeding almost exclusively on bamboo, the balance of the ecosystem is very easily damaged and destroyed. An adult panda needs to eat 12–18 kilograms (26–40 pounds) of bamboo every day, so if something affects the bamboo, the pandas are in deep trouble. Whole areas of bamboo tend to mature, flower, and die at the same time, which means that the pandas then have to move on to another area. This can be a problem, because the natural habitat of the panda is disappearing in many places, so there is often nowhere safe to move on to.

Most herbivores eat a variety of different types of plants, so if one type of plant is threatened, they can simply eat more of the others. When animals are part of a food web, rather than a single food chain, they are more likely to survive and adapt to changes in conditions.

SCIENCE PIONEERS Charles Elton again!

The young British biologist Charles Elton didn't only come up with the idea of food chains as he made his observations on Bear Island (see page 17). Almost as soon as he had come up with the idea, he recognized that a food chain was a very simplified version of what was really happening. Elton watched as Arctic foxes ate a wide range of animals on the island, not just tundra birds such as ptarmigans. In the winter, when food was scarce, they ate bits of dead seals left by another predator, the polar bear, and even polar bear dung! He saw that herbivores ate a wide variety of plants, and from his observations he created what he called a "food web". Charles Elton studied the relationships between the environment and the living organisms – what we would call the **ecology** of Bear Island.

A food web like this one seen by Charles Elton in the 1920s is still a simplification of the real situation, but it gives us quite a good picture of the way the living organisms on Bear Island interact together.

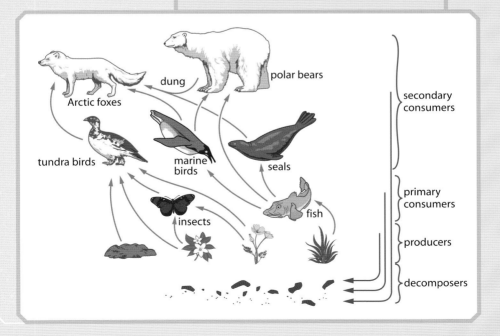

WEBS OF LIFE

Where do you find food webs? The answer is anywhere and everywhere. From deserts to mountainsides, from tropical rainforests to temperate woodlands, from sand dunes to rock pools, and from Arctic wastes to the depths of the ocean, living organisms thrive and feed on each other.

THE ANTARCTIC FOOD WEB

The Antarctic might not sound as if it is a very easy place to live, yet it is teeming with life. The waters of the ocean and indeed the ice itself are full of phytoplankton, bacteria, and protists, which are in turn eaten by microscopic zooplankton and **krill**. Krill are tiny shrimp-like crustaceans, and are the main food supply for the rest of the web. It has been estimated that the Antarctic phytoplankton feeds between 750 and 1,350 million tonnes of krill every year! The krill are then eaten by a wide variety of animals ranging from huge baleen whales like the blue whale and the humpback whale, to squid, fish, octopuses, crabs, and molluscs. Toothed whales, penguins, seals, and birds such as albatrosses and penguins are the secondary and tertiary consumers. They help to make a massive web of life covering thousands of square miles in the southernmost parts of the world.

Adult Antarctic krill measure only about 6 centimetres (2.4 inches) in length and weigh about 1 gram (0.04 ounce). They are small, but they are a vital part of the Antarctic food web.

WEBS CLOSER TO HOME

Most of us will never visit the Antarctic – but there are other food webs all around us wherever we look. A walk in a local woodland or park, or even just your own garden, takes you into the middle of a food web as rich as any found in the Antarctic. Things are simply on a smaller scale. Grasses, trees, bushes, and flowers are the producers for the web. Insects, snails, slugs, earwigs, and hundreds of other mini-beasts as well as birds, mice, voles, and rabbits are the main primary consumers. They in turn are eaten by bigger birds such as magpies, or by birds of prey, foxes, mink, moles, stoats, and weasels. And the biggest of these carnivores, the foxes and the larger birds of prey, will also eat some of the smaller carnivores if they get the chance.

Whatever the details, a food web very similar to this one is going on somewhere near you.

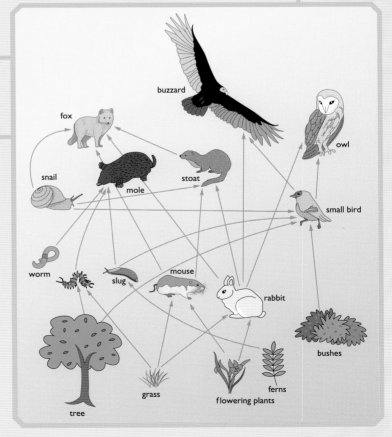

buzzard

fox

owl

snail

mole

stoat

small bird

worm

slug

mouse

rabbit

bushes

tree

grass

ferns

flowering plants

COMMUNITIES AND POPULATIONS

Knowing the feeding relationships (food chains and webs) between the different types of living organisms in a particular habitat isn't enough to let us really understand what is going on. Animals and plants live in groups of the same species, feeding and breeding together. These groups are called populations – the frogs in a local pond are a good example. Each pond has a population of frogs that live there, eating insects, and producing tadpoles to keep the population going.

In every habitat there are lots of populations of different types of organisms. The pond will also have a population of fish, a weed population growing on the bank, populations of water beetles, leeches, and many other types of plants and animals. All these populations together are known as a **community**. To understand how living things interact, we need to understand the way the different populations relate to each other within the community.

Your mattress is an example of a community. It is full of populations of dust mites, bacteria, and fungi all feeding on the dead skin cells you shed every night! They are all part of a food web that starts with the food you eat.

POPULATION CHANGES

Scientists study the numbers of animals or plants in a population, and watch how they change. The numbers go up as offspring are born or new organisms move in, and go down as organisms die or leave the area. In a stable population, the comings and goings more or less balance and the numbers stay roughly the same, but all sorts of things affect the size of a population, and these in turn can affect the food chains and webs in the community.

COMPETITION TO STAY ALIVE

The amount of food available, diseases, and new predators in an area all have an effect on population numbers. Competition between species is also a big factor. If there is a lot of competition for the food available, some types of animals will compete more successfully than others. The numbers of the populations that win the most food will go up, and the population numbers of the losers will go down. If there are lots of different types of animals all competing for the same food, then none of them will do very well.

It is not just animals that face competition – plants compete as well. They compete with other plants for light, space, water, and minerals and this affects how big they grow, and how many seeds they make.

When the numbers in the population of one type of organism in a food web go up or down, other populations in the community will be affected. There is only a certain amount of food and space in an area, so if one group of animals increases, they need more of the food and space. This leaves less food and space for the other animals in the community.

CLOSER LINKS THAN YOU THINK?

It is easy to see how a change in one population in a simple food chain would have an effect on the other populations involved. If eucalyptus trees died out, the koala bears that rely on them for food would probably die out as well.

But as we have seen, the situation in food webs is more complicated. Most animals eat a range of plants, and most carnivores eat a variety of animals. So the relationships between predators and prey may not be as simple as they seem. For example, some work in Canada on populations of snowshoe hares and lynxes gave a graph that looked very similar to the one for foxes and grouse below. The peaks and crashes of the lynx population and the hares appear to be linked just like animals in Scotland.

CLASSIC EXPERIMENTS The grouse/fox balance on Scottish grouse moors

The numbers of predators and prey in a community are often closely linked. This can be seen clearly by looking at the grouse and foxes on a moor, and how the numbers of each species change over time without interference from people. A simple explanation for this is that when there is plenty of plant food, the number of grouse increases. This means there is more food for the foxes, so more cubs survive and the number of foxes increases. But more foxes eat more grouse, so the numbers of birds fall. Then there is less food for the foxes, so the number of foxes falls again! This type of pattern can be found in many communities around the world.

The numbers of foxes and grouse in a community vary through the years but they remain linked together in a pattern.

Scientists showed that although the number of snowshoe hares really did have an effect on the numbers of lynx, the hare population went up and down in just the same way in habitats where there were no lynxes! Changes in the weather, insect pest populations, and parasites were affecting the food supply and health of the hares, and the population went up and down without any help from predators.

We still have a lot to learn about food webs. Research is going on all around the world to help us understand more about the ways in which animals and plants interact.

SCIENCE PIONEERS Dr Neo Martinez

Dr Neo Martinez and his teams at the University of California in Berkeley, US, and the Rocky Mountain Biological Laboratory, also in the US, have spent years studying food webs of different types, and in different places. They have built up amazing computer-generated images of food webs which they have used to show that most species in a community are very closely linked together. In fact, most of the species in the wide variety of habitats they looked at were only separated by two links of a food chain. This means that if one type of plant or animal becomes threatened, it can affect most of the other organisms in the community in some way. **Extinctions** and loss of **biodiversity** may have much bigger effects on a community of animals and plants than anyone had imagined.

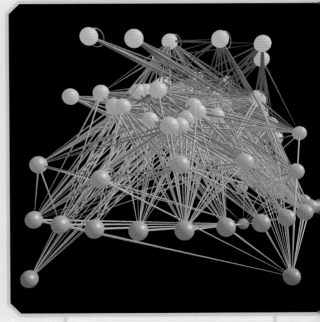

The amazing images of food webs from Neo Martinez and his team could help to change the way we understand the world around us. Each sphere represents a species, and each link represents a feeding relationship.

Cycles of life

Plants can make their own food, taking carbon dioxide from the air and water from the soil, and using energy from the Sun to turn them into simple sugars. But simple sugars are not enough for plants to live – they need to turn some of the sugar into proteins to build enzymes and cells. In order to make proteins, plants need minerals. They take these minerals from the soil through their roots. The proteins in the plant material are then eaten by animals and become part of their bodies in food chains and webs.

Eventually, the organism will die and dead plants, dead animals, and animal droppings are all then used as food by the decomposers. The carbohydrates, proteins, and fats that made up the living cells are broken down again into the elements that made them (such as carbon and nitrogen) and returned to the soil and the air. Without this recycling of resources, life would grind to a halt.

In a stable community of animals and plants, processes which remove minerals from the soil (for example, plants growing) are balanced by the processes of decay which put them back. In this way, materials such as nitrogen and carbon are constantly cycled through the environment.

THE CARBON CYCLE

The element carbon is vital for all living things, because all of the main molecules of life such as carbohydrates, proteins, fats, and **DNA (deoxyribonucleic acid)** are based on carbon atoms joined to other elements.

There is a huge pool of carbon for living things to use in the form of carbon dioxide. It is found in the air (it makes up about 0.04 percent of the air you breathe) and dissolved in the water of rivers, lakes, and seas. At the same time, carbon is constantly released back into the environment through respiration in plants and animals. During respiration, sugar is broken down in the cells using oxygen. This supplies the cells with energy and produces carbon dioxide and water as waste products. The work of the decomposers also releases all the carbon locked up in plant and animal bodies as they decay. This constant recycling of carbon in nature is known as the carbon cycle.

In a natural balanced ecosystem of animals and plants, the carbon cycle regulates itself. The oceans and forests can soak up huge amounts of carbon dioxide. However, in recent years humans have been adding more and more carbon dioxide into the atmosphere. Now there are fears that the natural cycle may be reaching breaking point.

This diagram shows the carbon cycle in nature.

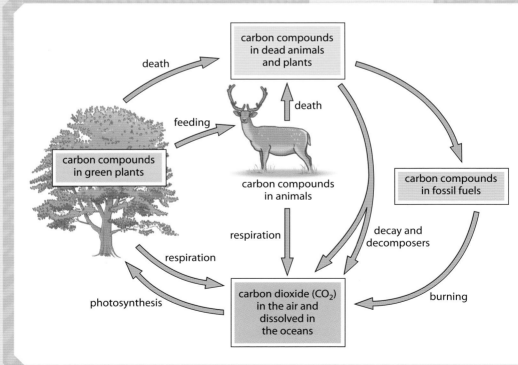

A CARBON CRISIS?

Over thousands of years, the levels of carbon dioxide in the atmosphere of the Earth have been fairly stable. But in the last fifty years or so, the amount of carbon dioxide we produce has gone up enormously. Burning **fossil fuels** to produce electricity, heat our homes, and power our cars releases carbon dioxide into the atmosphere, and we are burning more and more fossil fuels.

The seas and oceans hold about 98 percent of all the carbon available on and around the Earth because carbon dioxide dissolves in seawater. It dissolves more easily in very cold water than in warmer seas and oceans. As carbon dioxide levels in the air go up, more is absorbed by the seas and oceans. However, it takes time for the carbon dioxide to dissolve, and the seas cannot cope quickly enough with all the extra carbon dioxide we are producing. Scientists think that only about 40–50 percent of the carbon added to the atmosphere since 1800 has so far dissolved into our oceans.

Scientists have been measuring carbon dioxide levels on a mountain top in Hawaii for many years. The levels of carbon dioxide go up and down each year with the seasons – in summer the levels drop as the plants are growing fast and using lots of carbon, and in winter it rises again – but the overall trend is steadily upwards.

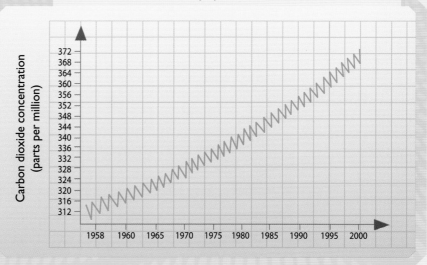

RECENT DEVELOPMENTS global warming

Many scientists believe that the increasing levels of carbon dioxide are adding to an effect called global warming or the greenhouse effect. The carbon dioxide acts like an insulating blanket around the Earth, keeping the heat from the Sun in. Many scientists also believe that if levels of carbon dioxide keep rising, the climate of the Earth will be changed drastically. This could lead to the polar ice caps melting which would cause big rises in sea levels. No-one is completely sure what is happening, but many countries are trying to reduce the amount of carbon dioxide they produce.

SCIENCE PIONEERS
James Lovelock and the Gaia theory

James Lovelock is a British scientist who has also spent many years in the US. He developed a theory that the Earth itself acts as a living organism on a very large and slow scale, and that conditions on the Earth will always return to those which make life possible. If people work as part of Gaia without interrupting the natural cycles, then all will be well. If we pollute the world too much, it will eventually right itself – but on a timescale which may be too late for us. Lovelock developed a computer programme which modelled his early theories of Gaia. As some of his predictions have already happened, his ideas have become more widely accepted.

The Earth as seen from space – the inspiration for James Lovelock's Gaia theory.

THE NITROGEN CYCLE IN NATURE

The carbon cycle is not the only system that plays an important part in maintaining life on Earth. The nitrogen cycle is also vital for living organisms. Nitrogen is a gas that makes up about 80 percent of the air we breathe. Plants also need nitrogen to make proteins and grow successfully. However, plants cannot use the nitrogen in the air. Instead they get it from the soil. They take nitrogen in through their roots as soluble **nitrates** and use it to build new plant material. When animals eat the plants, the nitrates in the plant protein become part of the animals' bodies.

Nitrates get put back into the soil in a number of ways. Urine contains a chemical called urea that is made as proteins are broken down in the body. Faeces contain lots of protein too, so animal waste products (including human ones) are a good source of nitrates for the soil. In the same way, when plants and animals die, their bodies (which contain a lot of protein) can be broken down by the decomposers to release nitrates.

Finally, there is another way for some plants (like beans and peas) to get nitrogen. They have special bacteria called *Rhizobium* which live on their roots. These bacteria can capture nitrogen from the air and turn it into nitrates which the plants are able to use. They are known as nitrogen-fixing plants. This is a very close feeding relationship. The bacteria get a safe place to live and a supply of sugars from the plant, while the plant gains a constant source of nitrates.

BALANCING THE BOOKS

In the wild, when a plant dies, it falls to the ground and the decomposers release the nitrates back into the soil as the plant rots. But when people harvest plants for food, the nitrates in those plants are not returned to the soil. For the soil to remain fertile, the nitrates need to be replaced in some way. This is usually done by adding fertilizers. In many of the developing countries, animal manure, human manure, and compost made from rotted plants are the main types of fertilizer used. They have to be broken down by decomposers before they release their nitrates into the soil.

In developed countries, chemical fertilizers are often chosen. They are made in huge quantities, in chemical processes which take nitrogen from the air and turn it into a form that can easily be taken up by plants. These fertilizers have made a big difference to the amount of crops that can be grown, and mean the fertility of the soil is no longer completely dependent on the nitrogen cycle.

This is the nitrogen cycle, on which all life depends.

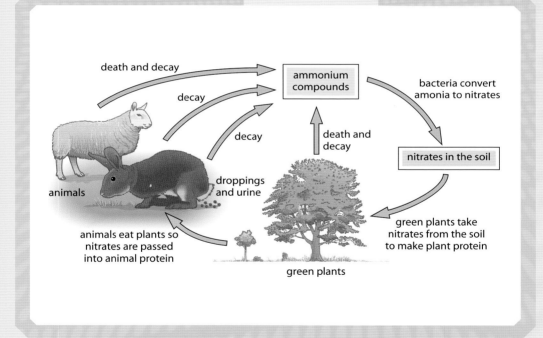

death and decay

decay

ammonium compounds

bacteria convert amonia to nitrates

decay

death and decay

nitrates in the soil

droppings and urine

animals

animals eat plants so nitrates are passed into animal protein

green plants take nitrates from the soil to make plant protein

green plants

A weak link in the chain?

The links between species in food chains and webs means that any change or interference can have serious effects. Sometimes a change in a food web comes about through completely natural causes but, all too often, the change that triggers a problem is the direct result of something people have done.

INTRODUCING SOMETHING NEW

As people travel the world, they carry different organisms with them. Sometimes introducing a new animal or plant to an area has benefits but, all too often, it causes problems. When Thomas Austin brought 26 rabbits from England to Australia in 1859, he only wanted to introduce a bit of sporting shooting on his estate. But the rabbits

Rabbits eat huge amounts and reproduce very quickly – three or four litters a year. The rufous hare-wallaby (pictured here) which usually only produces one baby per year simply cannot compete!

bred rapidly and ever since have been a major problem. Even today, about 90 million Australian dollars are lost every year, as rabbits eat grass that could otherwise feed sheep, and another 20 million dollars is spent on controlling them. The rabbits have driven several native Australian animals to the verge of extinction. The rufous hare-wallaby, the bilby, and the burrowing bettong are just three of the animals that have almost disappeared in Australia, along with a number of plants.

In an attempt to solve the problem, a rabbit disease called myxomatosis was introduced in the 1950s. Although the disease helped control the rabbit population, it was not completely successful. The rabbits have become more and more immune to the disease. Now Australians are trying a new disease, called calicivirus, in another attempt to get rid of their rabbit pest.

The prickly pear cactus also made a bid to take over the Australian countryside, and nearly succeeded, because none of the animals there could eat it. It took the introduction of a moth species with very hungry caterpillars to bring the situation under control (see page 56 for more about biological pest control).

HEDGEHOG HORROR

In 1974, a small number of hedgehogs were introduced on a Scottish island by a gardener who thought they would help control garden pests. They did, but unfortunately it soon became clear that they were also eating the eggs of the rare wading birds which nested on the islands in their thousands. The hedgehogs bred very successfully, and by 2002 the numbers of some species of birds had dropped by 60 percent. On nearby islands without hedgehogs, bird numbers stayed the same. The only way to stop the destruction of the birds was to get rid of all the hedgehogs. Some of them were captured and killed while others were rehomed on the mainland.

POISON!

Weeds are a problem for all farmers – they grow and compete with crops for water, light, and nutrients. Various animal pests, bacteria, and fungi can also attack a crop and feed on it. Over the years, farmers have come to rely on chemical **herbicides** and **pesticides**. Herbicides kill weeds but leave the crop plants unharmed, while pesticides are designed to kill the insects that might attack and destroy the crop.

However, weedkillers and pesticides are both poisons. When they are sprayed onto crops, they can get into the soil and be washed into rivers and streams, where they can affect other animals and plants. Another problem is that some poisons can become part of food chains. Even if the insects or small animals that eat the sprayed crops are not damaged themselves, the poison can stay in their bodies and be passed on to any animals that eat them. As these animals eat lots of the insects, the amount of poison in their bodies can build up to dangerous levels. It can then be passed on up the food chain and may cause serious damage or even death to the top predators.

CLASSIC EXPERIMENT research into the effects of DDT on top carnivores

DDT is a very effective pesticide, but it can build up in the body fat of animals that take it in. Not long after farmers started using DDT (in the 1960s), the numbers of big fish-eating birds like herons and ospreys started to fall. Scientists analysed the dead birds, and discovered that they contained high levels of DDT – not enough to kill the birds, but high enough to affect reproduction. The scientists found that the DDT-affected birds were laying eggs with very thin shells, which broke easily. This meant fewer young birds were successfully hatching and growing up, and so the overall numbers of the birds were falling.

Herons and ospreys do not eat crops, so how did the DDT get into their tissues? Research showed that the DDT sprayed on the fields was also being washed into rivers, streams, and ponds where it was taken up by zooplankton. The zooplankton were eaten by very small fish, who in turn were eaten by larger fish. The levels of DDT in each type of organism got higher and higher up the food chain. Finally the heron, as top carnivore, ate the larger fish and took on a damaging dose of poison. As a result of problems like these, the use of DDT was banned in many countries.

A food chain of disaster – fortunately scientists realized what was happening before all the herons and ospreys were lost.

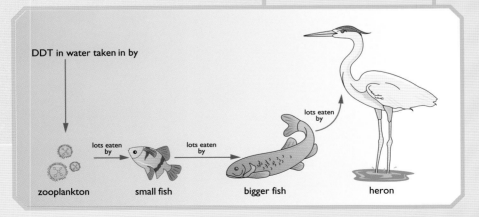

DDT in water taken in by

zooplankton — lots eaten by → small fish — lots eaten by → bigger fish — lots eaten by → heron

PAYING THE PRICE?

Problems in food chains and webs can affect human beings just as much as any other animal. We rely on industrial processes to supply us with many of the things we want and need, but sometimes the waste produced by these processes can cause problems. In the 1950s, a terrible disease appeared in the Japanese city of Minamata. Pet cats and people began to lose their balance. The people and animals affected first showed signs of muscle wasting, then they became paralysed and eventually died. Many babies were born suffering terrible deformities. Eventually the cause of the so-called "illness" was found.

There was a factory in Minamata that produced chlorine, sodium hydroxide, and hydrogen from seawater, using mercury

in the process. The factory waste, which contained some of this highly poisonous mercury, was released into the local seawater. It entered the local food chains and became more and more concentrated in the bigger fish, which were caught and eaten by the local people – and their pet cats. The mercury caused all the health problems that the people and cats suffered.

> If people had thought about food chains, the tragedy of Minamata need never have happened. The lessons learnt in Japan mean that people now think very carefully about the impact of industrial processes on the environment around them.

COLLAPSING WEBS

All over the world, natural environments are changing. Huge areas of rainforest are chopped down every day to make way for crops or cattle, often to provide cheap food for people in the developed world. Huge amounts of sewage and chemical waste are dumped into the seas and oceans of the world. Pesticides and herbicides are sprayed on to crops as we try to produce more food as cheaply as possible. We burn fossil fuels and pump carbon dioxide and other chemicals into the atmosphere faster than ever before.

At the same time, animals and plants are becoming extinct at a rate never before recorded in the history of the Earth. As one organism disappears, food chains and webs collapse and more animals and plants are put at risk or lost. Many scientists feel that there are strong links between human activities and extinction rates and predict serious problems ahead if things do not change.

BUILDING FOR THE FUTURE

However, there is still hope for the future. In many places, people are working hard to help preserve animals and plants, protecting food chains and webs and the environments that support them. People are managing the oceans and trying to avoid over-fishing. By controlling both what we put into our seas and what we take out, many countries are working together to keep the oceans of the world healthy for the future.

SCIENCE PIONEERS Aldo Leopold

Aldo Leopold (1887–1948) studied forestry at Yale University in the US, and worked in forestry and game management. He began to understand that the living things in a habitat all rely on each other. Aldo suggested that people needed to learn to "think like a mountain" – in other words, to think about all the different organisms in a habitat, and take the long view.

SUSTAINABILITY

Globally, there is a strong move towards farming in a way that looks after the land. "Sustainable agriculture" means taking care of our resources to make sure that they do not run out and that we replace what we use carefully. This can mean ploughing in the remains of crops, using animal waste as fertilizer, replanting hedgerows, and planting a mixture of trees in commercial woodlands. Meanwhile, people continue to investigate how farming methods can be improved both to produce more food and to protect the environment.

RECENT DEVELOPMENTS
biological pest control

Biological pest control uses one living organism to control another. For instance, you can now buy ladybirds to help control greenfly in your gardens. Using living organisms instead of chemicals to protect crops from pests reduces the need to use poisons, which means we are less likely to cause problems like the DDT disaster. However, as we saw with the Australian rabbits, introducing new species to an area can cause difficulties too. It needs lots of money and research to make sure that no unexpected problems arise.

There are several different types of biological controls:

- Biological pesticides. Sometimes a disease can be used to wipe out a pest animal. For example, *Bacillus thuringiensis* is a bacterium which can be sprayed on water or plants to be taken up by pest larvae. The bacteria make a poison that kills the larvae, but is harmless to anything else in the food chain. It has been used successfully in the US, UK, Europe, and Canada.

- Encouraging natural enemies. Making sure there are lots of the natural enemies of a pest may control it as well as any chemical. This sort of system is called integrated pest management. The government in Indonesia worked with scientists and rice farmers to control brown planthoppers, that can totally destroy a rice crop. The farmers were trained to identify the spiders which eat the planthoppers and conserve them. In just three years, the money spent on pesticides had dropped by 90 percent, and rice yields were increasing steadily.

- Exotic enemies. A species from one country can be introduced to control a pest in another country. A century ago, the citrus fruit growers of California were in despair as their trees were destroyed by an insect called the cottony cushion scale. The vedalia beetle from Australia was brought in to save the day.

In recent history there have been various problems with species introduction – see the prickly pear example on page 51 – which have nearly caused disasters. In the 21st century, a lot of care and research goes into the introduction of a new species. When it is used wisely, biological pest control is a powerful weapon.

The vedalia beetles devoured the scale insects, and so the citrus groves of California are still there today.

Managing the future

To help preserve life on Earth, we need to understand the feeding relationships between the plants and animals in all the different habitats of the world. Scientists in many different countries are working hard to bring us the information that we need. One example is the work of Dr Francis Gilbert (UK) and Professor Samy Zalat (Egypt). They have spent several years building up a detailed picture of the biodiversity of the Sinai region of Egypt. They have looked at all the living organisms in the area, from the smallest plants, through insects, birds, mammals, and even the parasites which infect everything else. Starting in 2005, the project is getting bigger. With funding from the **United Nations** they will try and record the biodiversity of the whole of Egypt.

Conditions in Sinai are extreme, yet a huge range of different living organisms make it their home, including spiny mice, ground beetles, spiders, and the smallest butterfly in the world, the Sinai Baton Blue.

ENGINEERING THE FUTURE?

In recent years, scientists have learnt how to change the **genetic information** in plants and animals. **Genetic modification** is being used in farming to add completely new **genes** to animals and plants. This is known as **GM technology**. GM crop plants can be made that produce more crops, grow to a larger size or have fruit that lasts longer. Sometimes new genes allow the plants to make chemicals that protect them from pests, so no spraying pesticide is needed. New genes can also allow the plants to make compounds that make them glow in the dark when they are attacked by pests, so the farmer only needs to spray the fields when it is really necessary.

However, this new technology is very controversial. No-one knows how GM crops might change the feeding relationships of the future. Many GM plants are also given genes to make them infertile, so farmers have to buy new seed each year. This can be a real problem in developing countries where farmers are very poor. Also, it is feared that if these genes get into wild populations of plants or animals, they could do real harm to food webs and chains, particularly if they make these plants infertile. In the UK and Europe many people are against GM foods, but GM crops such as corn, cotton, and soybeans have been produced in the US and other countries for several years now, and so far there are no signs of any of the problems people feared.

CHANGE FOR THE BETTER

The feeding relationships between living organisms are fascinating and complex. They are the basis for our study of ecology and our understanding of life on Earth. As we learn more about the interactions between different living organisms, perhaps we can also learn more about the ways in which human actions can affect the food chains and webs around us. Hopefully then we can make sure that when we change things, we make them better!

Further resources

MORE BOOKS TO READ

Solway, Andrew, *Wild Predators* series (Heinemann Library, 2005)

Stockley, Corinne, *The Usborne Illustrated Dictionary of Biology* (Usborne Publishing, 2005)

Wallace, Holly, *Life Processes: Food Chains and Webs* (Heinemann Library, 2001)

Nature Encyclopedia (Dorling Kindersley, 1998)

USING THE INTERNET

Explore the Internet to find out more about feeding relationships. You can use a search engine, such as www.yahooligans.com or www.google.com, and type in keywords such as *food chains*, *photosynthesis*, *consumers*, *Gaia theory*, or *nitrogen cycle*.

These search tips will help you find useful websites more quickly:

- Know exactly what you want to find out about first.
- Use only a few important keywords in a search, putting the most relevant words first.
- Be precise. Only use names of people, places, or things.

Disclaimer

All the internet addresses (URLs) given in this book were valid at the time of going to press. However, due to the dynamic nature of the Internet, some addresses may have changed, or sites may have ceased to exist since publication. While the author and publishers regret any inconvenience this may cause readers, no responsibility for any such changes can be accepted by either the author or the publishers.

Glossary

alga   simple plant

adaptation   special features of an organism that enable it to survive in a particular habitat

atmospheric pressure   value of air pressure in the atmosphere

bacteria   type of micro-organism that can be helpful, but that can also cause disease

biodiversity   measure the diversity of organisms living in a given area – both the different types of organisms and the variety within species

biomass   total mass of living organisms in an area

carbohydrate   type of food made up of carbon, hydrogen, and oxygen

carnivore   animal that eats only other animals

cell   small, simple building block of any living thing

chemosynthetic bacteria   bacteria which make food using the energy from other chemicals

chlorophyll   chemical used by plants to capture the Sun's energy

chloroplast   structure in the plant cell that contains chlorophyll

cnidocyte (also known as nematoblasts)   specialised stinging cell found on the tentacles of animals such as sea anemones

community   all the populations of animals and plants that live together in a habitat at any one time

culture   controlled growth in laboratory, for example bacteria

decomposer   organism that breaks down natural waste, and dead plants and animals

digest   break down into smaller molecules

digestive system   system made up of a set of organs that digest food into the molecules needed by the body

DNA (deoxyribonucleic acid)   molecule that carries the genetic code. It is found in the nucleus of the cell.

echo-location   working out distances using sound echoes, commonly used by animals such as dolphins, whales, and bats

ecology   scientific study of the ecosystem

ecosystem   all the animals and plants living in an area, along with the interactions between the living organisms and the things that affect them such as the soil and the weather

energy   capacity to do work

enzyme   protein molecule that changes the rate of chemical reactions in living things without being affected itself in the process

excrete   give off waste products

extinction   when a whole population of a species dies out completely

faeces   solid waste from the body

food chain   links between different animals that feed on each other and on plants

food web   model of a habitat showing how the animals and plants in different food chains are interconnected through their feeding habits

fossil fuels   fuels formed over millions of years from the remains of ancient plants and animals. They are oil, coal, and natural gas.

gene   unit of information in the DNA

genetic information   information contained within the chromosomes located in the nucleus of each cell

genetic modification   changing the genetic information in a cell

glucose   simple sugar made by photosynthesis and used as a fuel in the cells of the body

GM technology   using genetic modifictation for our benefit, such as growing crops that are resistant to pests

habitat   place where an animal or plant lives

herbicide   chemical that kills plants (usually used on weeds)

herbivore   animal that eats only plants

insulate   provide a barrier to hot or cold temperature

krill   shrimp-like crustaceans that are very important in marine food chains and webs

large intestine   part of the digestive system

lichen   simple plant that grows on rocks, walls, or trees

maggot   soft-bodied young form of an insect

malaria   fever caused by a protozoan parasite, more common in warmer countries

marine   to do with the seas and oceans

micro-organism   bacteria, viruses, and other minute organisms which can be seen only using a microscope

molecule   group of atoms bonded together

nitrate   mineral found in the soil. Nitrates provide nitrogen for plants.

omnivore   consumer that eats both plants and animals

organism   individual living thing, such as a plant or animal

pesticide   chemical that kills pests, which are often insects

photosynthesis   process by which green plants make food from carbon dioxide and water, using energy from the Sun

phytoplankton   micro-organisms found in the seas and oceans that make their food by photosynthesis

pod   group of whales

pollinate   transfer pollen from the male to the female parts of a flower

predator   animal that preys on other animals to obtain food

prey   animal that is preyed on and eaten by predators

primary consumer   first animal in the food chain. These are herbivores and omnivores.

producer   first organism in the food chain

protein   important building block of living things

protist   microscopic living organism, mainly single-celled

protozoa   single-celled microscopic animal

ptarmigan   ground-dwelling bird the size of a small chicken

pupa   young inactive insect in the process of becoming an adult

radioactive   gives off radiation

respiration   process involved in the production of energy in living things. Oxygen is taken in and carbon dioxide given off.

secondary consumer   second animal in the food chain. These animals eat the primary consumers.

sleeping sickness   disease caused by a parasitic protozoan, with effects including extreme tiredness

species   specific group of very closely related organisms whose members can breed successfully to produce fertile offspring

tertiary consumer   third animal in the food chain. These animals eat the secondary consumers.

trophic   relating to feeding or nutrition

trophic-dynamic model   model created by Raymond Lindeman to describe the trophic levels

tundra   treeless Arctic region of Europe, Asia, and North America

United Nations   worldwide association of governments

zooplankton   micro-organisms found in the seas and oceans that feed on phytoplankton

Index